50 Quantum Physics Questions In Plain Simple English Book 1

Simple and Easy Answers Without Math For Beginners

I0492604

Donald B. Grey

50 Quantum Physics Questions

Bluesource And Friends

This book is brought to you by Bluesource And Friends, a happy book publishing company.

Our motto is **"Happiness Within Pages"**

We promise to deliver amazing value to readers with our books.

We also appreciate honest book reviews from our readers.

Connect with us on our Facebook page www.facebook.com/bluesourceandfriends and stay tuned to our latest book promotions and free giveaways.

Contents

Introduction

Ever since the brilliant insights and accomplishments of Albert Einstein, Max Planck, Marie and Pierre Curie, Niels Bohr, Wolfgang Pauli, Enrico Fermi and their peers gained notoriety at the beginning of more than a century ago, the science of quantum physics has fascinated and tantalized a public hungry for knowledge, yet unable to decipher its mysteries and complexities.

As the first half of the 20th century unfolded, the discoveries of protons, neutrons, and electrons continued to evoke interest, which was further intensified when the atomic age began, demonstrating the awesome energy contained within the atom. Yet even today, many people envision a planetary model atom with a few "satellite" electrons orbiting close to a nucleus. Few can conceptualize a realistic image of an atom, or of its subatomic components, and how they interact with the governing forces that bind them. Are you wondering why theoretical physicists are fascinated by Schrödinger's cat?

This book is written in simple English for those who are interested in learning about quantum physics, but don't want to get into the mathematics, the formulas, and the deep complexities. You may have heard of the theories of relativity and uncertainty, or of quarks, strings, and superstrings, the strong force and the weak force, and want to understand them. You may be aware of electromagnetic energy, but don't know how it fits into quantum physics. You want to know about the pioneering scientists who created quantum physics and nuclear energy, and would like their theories explained in a way that you can go, "Aha, I get it now."

Rather than creating yet another textbook with chapter after chapter of ever-deepening details and with some formulas sneaking in, this book is in a format of 50 questions and their clear, brief answers. You can hop around to the questions that interest you most, but if you read it from beginning to end, you'll appreciate how we go from the Big Bang theory of the creation of the universe, and then on to the smallest and newest subatomic particles.

I hope that your questions will be well-answered, and that your curiosity will be stimulated to look further and explore the issues within quantum physics that interest you most.

Donald B. Grey

50 Quantum Physics Questions

Q 1. What is quantum physics?

Quantum physics, also called "quantum mechanics", is the branch of physics theory concerned with the smallest objects in the known universe, and includes atoms, the nucleus of the atom and its component protons and neutrons, the quarks within protons and neutrons, the electrons orbiting the nucleus, the particles and forces that hold the nucleus together— or cause it to split—and all of the strange, irrational behaviors that these particles and forces exhibit.

The term "quantum" derives from the Latin *quantus*, meaning "how much", and refers to Max Planck's concept that electromagnetic energy, including light, radio waves, electricity, and magnetism, radiates in discrete clusters, or quanta, and not as continuous physical fields or forces. The quantum particle that radiates light and other electromagnetic energies is called a "photon".

Quantum physics begins at the atomic level, then drills down into the atom's components and the forces that influence them. We begin with the atom intact and first look at its components, and then the forces.

Q 2. Where did the universe come from, and where is it going?

About 13.8 billion years ago, the matter and energy that exists today in the universe burst forth from an infinitely small point called a singularity. No one knows where all this matter and energy came from, although some theorists believe a previously existing universe had collapsed inwards and compressed through gravitational force to the singularity, then "bounced" back outwards. A similar theory postulates that a previously-existing universe fell into a supermassive black hole, which had a singularity at its center, and then exploded outward. Yet others suggest that our universe existed previously in another dimension of time and space, and "leaked" through to our dimension. Another concept, the multiverse, has our universe as one of many.

However it began, the universe underwent an extremely brief period of "inflation," or exponential growth, which smoothed out the expanding matter, then continued its expansion. At a certain point, the expansion of the universe inexplicably began to accelerate, and continues to do so today. Will the expansion continue indefinitely? That's another question, one that is not yet ready for an answer, although a force called "dark energy" is a leading candidate.

50 Quantum Physics Questions

Q 3. What is the cosmic microwave background, and why does it matter?

Physicists had been searching for tangible evidence of the primordial Big Bang and early expansion of the universe, believing that the initial outpouring of mass and energy had to leave a trace or echo. This was calculated to become visible 380,000 years after the Big Bang, when the hot universe cooled sufficiently to allow atoms to form in an event called "recombination" that caused a monumental flash of electromagnetic energy. At that moment, the universe went from opaque to transparent.

The search for the electromagnetic remnant, or echo, began in 1941, with preliminary measurements based on spectral lines, and culminated in 1965, when two radio astronomers, Arno Penzias and Robert Wilson, accidentally discovered the cosmic microwave background (CMB) while testing a new radio telescope that measured the microwave segment of the spectrum. This accomplishment earned them the 1978 Nobel Prize in Physics. Subsequent mapping of the CMB by dedicated satellites, notably the Planck space observatory, further refined its accuracy. The importance of the CMB includes its verification of the Big Bang theory as the origin of the universe.

Q 4. What are the three laws of thermodynamics?

The science of thermodynamics describes the transfer of thermal, or heat energy, into another form of energy. Despite their simplicity, these three laws place tight limits on how energy can exist in the universe.

The *first law of thermodynamics* is known as the Law of Conservation of Energy: the energy that goes into a system has to be fully accounted for, and cannot disappear; energy can neither be created nor destroyed. It may be converted into other forms of energy, or into matter, but all the energy that exists in the universe remains fixed and unchangeable. The energy of a system can change only when work or heat energy are added or removed. This first law is considered to be among the most all-encompassing scientific concepts.

The *second law of thermodynamics* is known popularly as the Law of Entropy, which is the disorder that takes place in a closed system. Over time, any system goes from ordered to disordered, from gas molecules in a container to stars and galaxies. Imagine a child's bedroom being cleaned and organized, and then what happens to it over time. The entropy of the universe increases over time, and continues to do so. Heat and energy can freely move to lower energy levels, but cannot move to higher levels without additional heat or energy being supplied.

The *third law of thermodynamics* concerns the behavior of particles at extremely low temperatures, with all motion stopping at Absolute or Kelvin (K) zero. As temperature rises above zero K, particle motion moves from the ordered non-motion state at zero K and becomes increasingly random, a condition we know from the second law as "entropy". According to the third law, entropy and temperature are equivalent.

Q 5. Will we disappear into a black hole?

Within a relatively brief time (a few decades), black holes have gone from a theoretical concept to a confirmed reality. The existence and size (mass) of black holes has been inferred from the tremendous gravitational forces they exert. More recently, black holes have actually been "seen," as matter being pulled, or accreted, into a black hole emits vast amounts of electromagnetic energy, visible to radio telescopes. It is now believed that quasars—the large, ultrabright objects at the far edge of the visible universe, are actually the accretion disks of matter being pulled into massive blacks holes.

A black hole is the remnant of a collapsed massive star, with a gravitational force at its center so powerful that even light cannot escape. Black holes grow in mass as they absorb nearby stars and other matter.

You don't have to worry about getting pulled into a black hole, since there are none nearby. A giant black hole, with the mass of millions of stars, is at the center of our Milky Way galaxy, but that's about 25,000 lightyears, or 125 quadrillion miles, from here.

50 Quantum Physics Questions

Q 6. What is dark matter?

Much of what is learned in physics is discovered by inference, such as the black holes we just discussed. An example of a current inference is the concept of dark matter, which is believed to account for 27% of the mass of the universe. That's quite a bit of material, but we can't see it, and don't yet know what it is. The existence of dark matter is based on the observations of astronomers that stars rotating in galaxies, especially at the outer edges, should be trailing stars closer to the center due to the greater distance they have to travel. Some unseen, huge amount of mass is creating a gravitational attraction that holds all the stars in galaxies in tight orbital positions. Here's where dark matter comes in:

Quantum physicists theorize that dark matter is a subatomic particle that exerts sufficient gravitational force to account for the rotational patterns of stars. Obviously, there has to be an incredible amount of these dark matter particles throughout the universe to exert this effect. Recent theories suggest the neutrino, which is almost, but not quite massless, to be dark matter: What the neutrino lacks in mass individually, it makes up for collectively. Trillions pass through our bodies every second. Others propose some yet-unidentified exotic particles comprise of dark matter. The investigation continues.

Q 7. Is dark energy an unknown new force?

Early in the 20th century, Albert Einstein proposed a hidden force in the universe called the "cosmological constant," which permeates space and contributes to the expansion of the universe. He later recanted his theory, with regrets. However, it now appears that he may have been right by predicting what we now call "dark energy", which accounts for about 68% of the mass of the universe, more than double the mass of dark matter, and far beyond the remaining 5% attributable to the visible universe.

The expansion of the universe was first observed by astronomer Edwin Hubble in the 1930s, but even more surprising were recent determinations that the expansion is accelerating. In 1998, Hubble Space Telescope studies confirmed that the expansion began to accelerate about six billion years ago, based on the positions and distances of supernovae (exploding stars), an effect that defies the fundamental force of gravity.

The yet-unidentified dark energy was thought to be a repulsive, negative force of gravity that pushes matter apart. Was Einstein correct? The search for answers continues.

Q 8. Is everything made of atoms?

Yes, if we include everything we call "matter". The ancient Greeks used the word "atom," which means "indivisible", to describe the point at which we can no longer divide matter. As far as science knew, this was an accurate description until the late 1930s, when atoms were split and the concept of atomic energy became a reality. It's correct to say that all matter is made of elements, or of molecules that are made of elements, and every element is made of atoms.

There are 92 naturally-occurring elements, each with its own unique number of protons contained in the nucleus, from number one, hydrogen, the lightest, to number 92, uranium. (In addition, new elements have been created in particle accelerators by adding protons to heavy elements; to date, there are 26 man-made elements, bringing the total to 118.)

Normally, for every proton in its nucleus, an atom has an equal number of electrons orbiting the nucleus. The positive charge of one proton is balanced by the negative charge of one electron, so an atom has a neutral charge unless it gains or loses electrons when it becomes an ion, or charged atom.

50 Quantum Physics Questions

Q 9. What is an isotope?

Except for hydrogen, every atom typically has a number of neutrons in its nucleus. Smaller elements have equal numbers of protons and neutrons; for example, carbon, with six protons, typically has six neutrons, and larger elements tend to have more neutrons than protons, up to the heaviest, uranium, with 92 protons and 146 neutrons.

There are exceptions: atoms can have a different number of neutrons than their average; hydrogen can add a neutron to become the hydrogen isotope deuterium, or two neutrons, to become tritium. Carbon can add 2 additional neutrons to become carbon 14, which is used in carbon dating. At the other extreme, 0.72% of naturally-occurring uranium atoms have 143 neutrons; known as uranium-235 (vs. normal uranium-238), this isotope is radioactive. Enriched uranium is refined to raise the concentration of uranium-235 to 5% or more for use in nuclear reactors.

Q 10. What's the difference between protons and neutrons?

As you have seen in the previous questions, the nucleus of the atom is made entirely of protons and neutrons. Except for electrical charge, they are virtually identical in mass, and their combined number equals the mass of an atom. A helium atom with two protons and two neutrons has an atomic mass of four (the two protons give it an atomic number of two). Carbon-12 (the normal form) has six protons and six neutrons, and the isotope carbon-14 has eight neutrons. (See above for the uranium example.) An electron has about 1/2,000 the mass of a proton or neutron, and is not counted in the atomic mass.

The one difference between protons and neutrons is the charge. One proton has a positive electrical charge, which exactly opposes the negative charge of one electron. The neutron, as its name implies, carries no electrical charge.

You may be wondering how a number of positively-charged protons stay tightly-held in the nucleus, and why the neutron remains there as well. That will be answered when we deal with fields and forces.

50 Quantum Physics Questions

Q 11. How do electrons stay in their orbits?

The configuration of the atom evolved from earlier concepts, including planetary models with close-orbiting electron "planets." Today we know the electron encircles the nucleusin multiple layers or levels called shells, orbitals, and suborbitals. These orbits are far out, creating a vast empty space between the nucleus and its electrons. If the nucleus were one inch in diameter, the nearest electron would be 200 feet away, in rough approximation. Another analogy suggests a basketball-sized nucleus having its electrons two miles away.

Newer theories describe electrons not as particles, but as waves of electromagnetic energy surrounding the nucleus as a cloud. They are not rotating, but are in undetermined positions, according to the uncertainties of quantum mechanics. The energy and electrical charges of electrons are well understood, but their positions and momentums are speculative.

50 Quantum Physics Questions

Q 12. Are protons and neutrons made of quarks?

Quarks are fermions, which include subatomic particles with mass. Both protons and neutrons, which are also fermions, contain quarks, but the quarks contribute only a small amount of the mass of these particles. The role of the quarks is to create the electromagnetic charge of the proton, and maintain the neutral electrical charge of the neutrons. Quarks were discovered in the 1960s by physicist Murray Gell-Mann.

There are three quarks in a proton: two "up quarks," each with ⅔ of one positive charge, and one "down quark" with ⅓ of one negative charge. Combined, the net charge is +1 (+⅔ +⅔ -⅓ = 1). Thus, the proton has a charge of +1, equal and opposite to the -1 of one electron.

There are also three quarks in a neutron: one "up quark" with ⅔ of one positive charge, and two "down quarks," each with ⅓ of one negative charge. Combined, the net charge is 0 (+⅔ -⅓ -⅓ = 0). As a result, the neutron has neither a negative or positive charge.

Quarks can change their charge, turning protons into neutrons and vise-versa. This occurs during beta decay, which will be covered later.

50 Quantum Physics Questions

Q 13. What is the difference between our Sun and a neutron star?

The star that warms us, the Sun, is about 4.5 billion years old, and generates heat and light though the process of nuclear fusion, in which hydrogen atoms fuse into helium atoms, and the energy of fusion is released. The energy slowly works its way to the surface of the Sun, and then makes the 93-million mile journey across space to Earth, traveling as photons, in a little over eight minutes. The majority of atoms in the Sun are hydrogen, helium, and a smaller amount of light elements: lithium, beryllium, boron, and carbon.

A neutron star, as its name implies, is composed entirely or primarily of neutrons. Lacking electrons encircling the nucleus at a great distance, the neutrons are able to squeeze tightly together, forming a star of incredible density. A neutron star results from the collapse of a supermassive star, which leaves a core of neutrons of 1.5 or more solar masses, packed into an ultradense ball with a diameter of only 12 miles. Except for a black hole, a neutron star is the smallest, densest stellar body in existence.

Q 14. Does antimatter really exist?

Yes. Antimatter is not a mysterious intangible, but a real physical entity, with properties that are the exact opposite of normal matter. Let's recall that matter is composed of atoms, with positively-charged protons and neutrally-charged neutrons in the nucleus, which is encircled by negatively-charged electrons. But in an antimatter atom, things are reversed: the proton has a negative charge and is called an "antiproton", and the electron has a positive charge, and is called a "positron".

We don't tend to run into antimatter casually, as it only exists on Earth when produced at the Large Hadron Collider (LHC) and other particle accelerators, where atoms are smashed into their component parts. As the energy of the early universe cooled and coalesced into atoms, both matter and antimatter were produced. When they meet, atoms of matter and antimatter annihilate each other, and a leading theory holds that after all the antimatter in the universe was annihilated, enough matter remained to form the vast gas nebula that condensed into the stars.

Q 15. Did Marie & Pierre Curie discover radioactivity?

Radioactivity had already been discovered in 1896 by Henri Bequerel in France, when he placed uranium on a photographic plate. The plate was wrapped in black paper, yet became exposed by the radiation emitted by the uranium. This was the first proof that atoms can change and be unstable, and an element can change to another element, with a different number of protons. The year before, 1895, Wilhelm Roentgen discovered x-rays, a form of radiation.

The husband and wife team of Pierre and Marie Curie (who was Bequerel's graduate student) worked for several years in a Paris laboratory to extract the highly radioactive element radium from an ore called "pitchblende". The Curies also discovered the radioactive elements thorium and polonium (named in honor of Marie's home country). The Curies received the Nobel Prize in Physics, with Marie being the first female recipient, and later the first person to receive two Nobel Prizes.

50 Quantum Physics Questions

Q 16. Is nuclear fission the same as splitting the atom?

Nuclear fission is the splitting of a large atomic nucleus into two smaller atoms. This may occur spontaneously, due to an imbalance among the protons and neutrons, or as the result of a neutron striking the nucleus of a radioactive element, like the isotope uranium-235. In either case, an imbalance occurs between the electrostatic repulsion of the positively-charged protons, and the strong nuclear force which binds the nucleus. For example, a stray neutron, turned loose by a uranium-235 atom, may strike the nucleus of another uranium-235 atom (with its 92 protons), and cause it to split into two different elements, typically barium, with 56 protons, and krypton, with 36 protons (notice how these add up exactly to 92 protons).

Unlike the "parent" uranium-235 atom, the two "daughter" atoms are stable, and not radioactive. When nuclear fission occurs, the combined mass of the two daughter elements is slightly less than the mass of the parent element. This is due to two neutrons being released during fission (and free to start fission in other atoms), and the conversion of mass to energy during the release of the binding energy that holds the nucleus together. This energy is substantial: in aggregate, one kilogram (2.2 lbs) of uranium-235 contains the heat energy equivalent of 4-billion kilograms of coal.

Q 17. Is alpha decay the same as radioactive decay?

In a spontaneous process called "alpha decay", that occurs only among radioactive elements, an atom of a radioactive element, such as uranium-235, may fire off an alpha particle, which is made of two protons and two neutrons, and which you may recognize as the nucleus of element number two, helium. The original uranium atom that fired off the alpha particle is now left with 90 protons, converting it to the element thorium. In turn, thorium can experience alpha decay, losing two protons and two neutrons to become radium, with 88 protons. Radium then may decay to radon, with 86 protons, and which then can decay to polonium, with 84 protons. Polonium then makes one final alpha decay to lead, which is a stable element with 82 protons, and the cycles of alpha decay are over. All lead that exists on Earth was originally a heavier radioactive element that gradually decayed.

About 99% of the helium generated on Earth is the result of alpha decay taking place underground. The helium that escapes to the surface rises into the atmosphere, which is why helium was first discovered in the spectrum of sunlight; hence the name, from the Greek "helios," for representing the sun god.

50 Quantum Physics Questions

Q 18. What is meant by "beta decay"?

While alpha decay involves conversion of elements by releasing of protons, beta decay involves electrons in the conversion of neutrons into protons, and protons into neutrons. This process involves the subatomic particles, quarks, which determine electrostatic charges, and includes emissions of ultralight neutrinos.

The neutral charge of a neutron is based on it containing two negatively-charged "down" quarks and one positively-charged "up" quark. The two $-\frac{1}{3}$ down charges are balanced by the one $+\frac{2}{3}$ up charge (see Q12). During beta decay of a neutron, one of the two negatively-charged down quarks changes to a positively-charged up quark, resulting in a $+1$ charge; the neutron has become a proton. In the nucleus of an element, the number of protons would increase by one (e.g. 92 uranium to 93 neptunium).

Two important changes occur at this time to complete the conversion of neutron to proton: An electron is emitted and takes an orbital position to electrically balance the positively-charged proton, and an antimatter neutrino is emitted into space. At this moment, the newly-formed proton is mathematically correct in mass and charge.

In the reverse conversion from proton to neutron, one of the two positively-charged up quarks converts to a negatively-charged down quark. Now two -⅓ down charges are balanced by one +⅔ up charge, and the proton is now a neutrally-charged neutron. Instead of an electron being emitted, a positively-charged positron is emitted, along with a neutrino. In the nucleus of an element, the number of protons would decrease by one (e.g. 92 uranium to 91 protactinium).

50 Quantum Physics Questions

Q 19. Does radioactivity release gamma rays?

When an atom emits an alpha particle (helium nucleus) or splits during nuclear fission, gamma rays are emitted. These are ultra-high frequency energies or particles, transmitted as waves with the shortest wavelength of any electromagnetic radiation. Unlike alpha and beta particles, gamma rays have no charge and no mass. Given their very short wavelength and high frequency, gamma rays are often considered similar to x-rays, which are also of very high energy, but gamma rays are even more energetic. They are capable of passing through any material, but are easily deflected by atomic nuclei, which can absorb their energy.

While gamma rays on Earth are created during nuclear fission or alpha decay, huge amounts of gamma radiation are created in space by large catastrophic events, like exploding stars (supernovae) or a collision of stars. When these bursts reach Earth, they do not cause dangers to people because most of the gamma radiation is absorbed by the atmosphere.

50 Quantum Physics Questions

Q 20. What is the half-life of a radioactive element?

The half-life of an unstable isotope of an element is the amount of time required for the isotope to emit, or decay, half of its radioactive atoms. Radiation, including the emission of an alpha particle, or the spontaneous splitting of an atomic nucleus, occurs on a random basis, since according to quantum theory, it is not possible to know exactly which atoms will decay, or when. But radiation is also predictive, as it can be determined how many atoms of an unstable element will decay during a given time. Once the half-life of an element is known, it remains unchanged and consistent over time. Half of whatever remains will decay during the half-life. An element with a short half-life of 20 days will decay half of its unstable atoms in 20 days, and then another half (of what remains) during the next 20 days.

Some radioactive elements have a long half-life: Uranium-235 has a half-life of about 7,000 years, radium has a half-life of 1,601 years, and yet the half-life of some isotopes can be as short as hours or even minutes. A radioactive isotope of technetium used in PET scans has a half-life of a few hours, so that the radiation does not remain long within the patient.

Q 21. What is nuclear fusion and where does it take place?

Nuclear fusion takes place primarily within the core of stars, where the tremendous gravitational pressure generates temperatures of 10 million (F) or more, and forces hydrogen atoms tightly together to fuse first into deuterium isotopes and then into helium atoms. This causes the energy of fusion to be released. This energy, as with nuclear fission, *only far greater*, is created by the release of some of the binding energy that holds the nucleus together.

The majority of atoms in the Sun are hydrogen, helium, and a smaller amount of light elements: Lithium, beryllium, boron, and carbon. In larger stars, nuclear fusion can create heavier elements, all the way up to iron, with 28 protons. Very heavy elements, all the way up to uranium, can be created during supernova explosions, when a star, having exhausted most of its fuel, explodes and generates even greater levels of heat and pressure.

The only nuclear fusion to take place on Earth has been the result of hydrogen bomb thermonuclear explosions, when the intense heat of an atomic bomb initiates nuclear fusion in a surrounding layer of deuterium atoms. Efforts are underway to initiate "cold fusion" for peacetime energy uses, but to date, this goal is proving to be elusive.

50 Quantum Physics Questions

Q 22. What are the forces included in the Standard Model?

There are four known forces that govern all interactions of subatomic particles and matter in the universe: Electromagnetism, the weak force, the strong force, and gravity. The first three are included in the Standard Model; the fourth, gravity, is not. The Standard Model combines two theories: The electroweak theory, which includes both the electromagnetic force and the weak force that mediates beta decay, and quantum chromodynamics, which describes the strong nuclear force, which holds protons, neutrons, and quarks together in the nucleus of atoms. These theories are called "gauge field" theories, meaning they are based on messenger, or force-carrying particles.

For electromagnetic forces, including light, electricity, magnetism, radio, x-rays, microwaves, and gamma rays, the messenger particle is the photon. For the weak force, the Z and W bosons mediate as particles, and for the strong binding force, the messenger particle is the gluon.

A messenger particle for gravity, hypothesized to be the 'graviton,' has not yet been discovered, thus its exclusion from the Standard Model.

Q 23. Do subatomic particles have angular momentum and spin?

Angular momentum and spin define the revolving and rotational motion of objects, from subatomic particles to massive stars and galaxies, and are based on the mass of the object and its momentum. On the macro scale, for illustration, the Earth's rotation around the sun has orbital angular momentum, and its rotation daily on its axis gives it spin angular momentum. The magnitude of angular momentum is based on linear momentum, which involves its mass and velocity, times the distance from the center point of its distance to the center of rotation.

Now, what is angular momentum and spin at the quantum, subatomic level? An electron's angular momentum is based on its orbital motion around the atom's nucleus, and the spin of the electron creates its spin angular momentum. But electrons are not solid objects, and according to quantum laws, they are points rather than spheres, and their exact position is uncertain. What is known is that, unlike a baseball, whose angular momentum and spin will vary, especially when affected by gravity, the momentum and spin of subatomic particles is steady and unchanging. The electron is determined to have ½ integer spin.

50 Quantum Physics Questions

Q 24. What is the smallest-sized unit of energy?

Telescoping down to the smallest energy units brings us to the photon and the Planck constant. The photon is the particle that transmits, or carries, one quantized unit of electromagnetic energy, and is considered the smallest unit of energy. At its absolutely lowest energy level, a photon carries a charge of one eV, or one electron volt. This is the amount of energy an electron needs to "jump" from one orbital energy level to a higher level.

When an electron does the reverse, and descends to one lower orbital level—it releases one eV and emits a photon of light. For the light to be in the frequency of red, more energy, 2.48 eV, would be needed, which is a descent of about three orbital levels. Physicists measure energy at this level using the Planck constant, the smallest amount of action (energy * time), which is represented in formulas by h; it is an almost infinitesimal fraction of a joule, approximately $6.626 * 10^{-34}$ (a joule is the amount of energy needed to move one kilogram a distance of one meter).

Q 25. Are atoms made of fermions or baryons?

Named after Nobel laureate Enrico Fermi, who created the first nuclear chain reaction, fermions include the most fundamental particles with mass: Electrons and antimatter positrons, neutrinos and antimatter neutrinos, muons (like electrons, with one eV, negative charge, and ½ spin, but 200 times the mass), and quarks. Protons and neutrons, each of which are made of three quarks, are classified as composite fermions, so yes, the atom and its components are made of fermions. In contrast, a photon of light or other electromagnetic energy has no mass, and therefore is not a fermion.

A baryon is a larger category of particles, such as protons and neutrons and lesser known particles, all of which are made of three quarks. So, protons and neutrons may be called "baryons" or "composite fermions". Baryons, in turn, are part of a larger class of particles called "hadrons", which are subject to the strong nuclear force holding them together in the nucleus.

Q 26. What is the difference between hadrons, baryons, and mesons?

"Hadron" is the overall term for subatomic particles that contain two or more quarks, which are held closely together by the strong nuclear force. It's similar to how atoms in molecules are held together by the electromagnetic force. Protons and neutrons are the most common form of hadrons. Antimatter baryons are antiprotons and antineutrons.

There are two families within the hadron family: Baryons, which contain odd numbers of quarks, and a category of particles containing an even number of quarks, called mesons. The typical meson, such as a pion, has one quark and one antimatter quark, or antiquark. A pion is an unstable, ultralight particle that may be formed when a cosmic ray collides with an atom in the atmosphere.

Leptons have small amounts of mass, but are not subject to the strong nuclear force; instead, they are affected by the weak force, and include electrons, muons, and neutrinos.

Q 27. Is a boson a particle with mass or energy?

Unlike fermions, baryons, hadrons, mesons, and leptons, bosons are subatomic particles that have no mass, yet can have effects on massive particles. Bosons include photons, the gluon, the $W^\pm$ and Z^0 weak forces, and the newly discovered "Higgs particle".

Photons can be either particles or waves (we'll cover that in another question), and are the force that mediates, or carries, light, electricity, magnetism, x-rays, and gamma rays. A photon is a quantum of energy—the smallest unit.

The gluon is the strong force which binds quarks within protons and neutrons, and binds protons and neutrons within the atomic nucleus. It is the strongest force in nature, yet only extends a very short distance, so disruptions in the nucleus can weaken its grip.

The weak interaction mediates beta decay, which involves conversions of particles within the atomic nucleus. This is covered in the next question, and we'll also cover the Higgs boson further down.

Q 28. How does the weak force mediate beta decay to change particles?

The beta force is called "weak" because it operates over an extremely short distance compared to the strong and electromagnetic forces; it is actually massive and more powerful than the electromagnetic force. There are two types of beta decay: The W⁻ and W⁺ weak forces, also called "weak interactions", that mediate beta decay in neutrons and protons.

Beta-*minus* decay by the negative weak interaction (W⁻) begins with a neutron. The interaction changes one of the neutron's two negatively-charged down quarks to a positively-charged up quark, and results in the emission of an electron and an antineutrino, which converts the neutron into a proton. This type of beta decay occurs generally within 10 minutes of a neutron being free of an atomic nucleus, but may occur within the nucleus as well under certain conditions.

Beta-*plus* decay is the reverse process: The positive weak interaction (W⁺) changes one of the proton's positively-charged up quarks to a negatively-charged down quark, and results in emission of a positron and a neutrino, neutralizing the proton's charge and converting it to a neutron. Beta-plus decay occurs less frequently in nature than negative decay, since protons are more stable than neutrons.

50 Quantum Physics Questions

The Z^0 weak force carries a neutral charge, and primarily affects the momentum, energy, and spin of electrons when they absorb energy from collisions with neutrinos, which have almost no mass but great momentum due to traveling close to the speed of light.

50 Quantum Physics Questions

Q 29. Why has the Higgs boson received so much attention?

Ever since it was theorized to exist in the 1960s by University of Edinburgh physicist Peter Higgs, there has been a search for what some call the "God Particle," a carrier particle of the hypothetical Higgs field, which, if it exists, permeates space and interacts with all subatomic particles in the universe and gives them mass.

The Standard Model does not provide for matter—quarks, electrons, protons, neutrons—to have mass; the Higgs boson was conceptualized to create a Higgs field—in effect, a fifth fundamental force that imparts mass to these particles, and all matter. In the world's largest particle accelerators, protons and atomic nuclei were smashed together, hoping to reveal previously-unknown ultramassive particles.

Proof of the Higgs particle's existence came in July 2012, when physicists at the Large Hadron Collider (LHC) in France/Switzerland reported discovery of a particle with a mass around 125-billion electron volts (125 giga eV), which was the hypothesized mass of the Higgs boson. Confirmation from additional experiments was reported in March 2013, culminating nearly 50 years of research and resulting in a Nobel Prize in Physics for professor Higgs and François

Englert, a Belgian physicist who collaborated with Higgs.

The Higgs is unique from other bosons in that it has momentum, but no direction, and no spin, and helps explain why photons have no mass, but the weak force W and Z bosons, which influence subatomic fermions, are massive.

Q 30. How can light be both a particle and a wave?

The concept of wave/particle duality resulted from speculation that if a wave must travel through a medium, as sound waves travel through the atmosphere, how can light, or other forms of electromagnetic radiation, travel through the empty vacuum of space? For centuries, it was believed that space was permeated by ether, a medium of propagation enabling waves to travel, but the existence of ether was later disproved, suggesting that light had to be a particle to travel through a vacuum.

Experimental physicists developed the Transactional Interpretation, stating that light moves from point to point as a wave, but where it is emitted or absorbed, light behaves like a particle. Two classical experiments verified the particle nature of light coexisting with its ability to be a wave, as shown in the following question.

Q 31. How was wave/particle duality proven?

Light is a form of energy, or force, within the electromagnetic spectrum. Like all forms of electromagnetic energy, it can be either a wave or a particle: It depends on the medium it is traveling through or reacting with, or the method used to measure it.

The photoelectric effect, in which incoming light striking metal can release electrons, was shown by Einstein in 1905 when he demonstrated that light's energy was contained in discrete quanta of energy, which we now recognize as the "photon", a particle.

The double-slit experiment, developed early in the 1800s, demonstrated that light passing through two parallel slits and striking a board forms alternating stripes, proving that light is a wave, as the waves of light cancel each other out through interference. But then a century or so later, when Einstein and others used detection devices to identify which waves of light passed through each slit, the overlapping waves changed to narrow slits of light, meaning the light was now passing through the slits as particles (photons), not waves.

Q 32. Does the uncertainty principle mean quantum physics can't measure things?

As quantum physics evolved through the first half of the 20th century, it became increasingly apparent that much of the logic of classical physics did not apply to the quantum world. One of the most direct manifestations of this is uncertainty—the mathematical inequalities that limit precise measurement of subatomic activity.

During meetings with his mentor, Niels Bohr (see following question), the German physicist Werner Heisenberg developed the uncertainty principle, which limits the precision of measurement of subatomic particles predicted from initial conditions. For example, knowing the momentum of a particle limits the ability to measure the particle's position, and vice-versa. Or, the more accurately you measure momentum, the less accurately you can measure position. Heisenberg's concept refuted the more logical-seeming model of the atom proposed by his Danish colleague, Niels Bohr, who envisioned electrons orbiting the nucleus in discrete shells; Heisenberg's uncertainty principle treats electrons as a diffuse cloud of electrical energy. In addition to pairs like momentum and position, the uncertainty principle also applies to time and energy, and angular position and angular momentum.

Q 33. The Copenhagen Interpretation of quantum mechanics

This was the culmination of informal, but inspired and creative meetings in Niels Bohr's institute in Copenhagen during the 1920s and 1930s, with colleagues Werner Heisenberg and others. The goal was to formalize quantum mechanics and move away from previous associations with the principles of classical physics, which does not concern atoms and subatomic particles, forces, and interactions.

In addition to Heisenberg's uncertainty principle (see the previous question), the meetings resulted in Bohr's complementarity principle, which states that at the atomic and subatomic levels, physical events may appear differently based on the nature of the observation. A key example is the wave/particle duality effect (see Q30), in which light can be either a wave or a particle.

Born's probability rule also emerged. Another participant at Copenhagen, the German physicist Max Born postulated in 1926 that there are probabilities that apply to every measurement, based on the square of the magnitude of a particle's wave function.

Interestingly, Einstein opposed the concept that particle physics is filled with imprecision, improbabilities, and uncertainty, and was quoted to say that "God does not play dice with the universe."

Q 34. Does "Schrödinger's Cat" experiment demonstrate superposition?

The previous questions on radiation, probabilities, and uncertainty are a good preparation for the explanation of physicist Edwin Schrödinger's celebrated cat experiment. He was being sarcastic about the improbabilities and uncertainties of quantum physics, and suggested the following scenario: A cat is placed in a box with a radioactive element and a poison. The conditions are such that the poison would be released if the element decays and emits an alpha particle (the assumption is that the radioactive element has a very short half-life, so there's a good likelihood that a particle will decay within a reasonable time).

At any time after the box is closed, the question is whether the cat is alive or dead. Only when the cover is lifted can the cat's state of existence be determined, since the cat's fate is dependent upon the uncertainty of radioactive decay. This demonstrates superposition, in which a particle can be in more than one place at a time. Schrödinger considered his cat experiment to be frivolous, but somehow it has stood the test of time and remains a popular lesson on subatomic improbabilities.

Q 35. What is Schrödinger's wave equation (and why is it important)?

Continuing our attention to Edwin Schrödinger, his wave equation is considered one of the founding concepts of quantum theory. It is defined as a differential equation that applies the Law of Conservation of Energy, which combines kinetic and potential energy levels to equal complete energy, and is used to learn about an electron's behavior relative to the nucleus of an atom. This is achieved by calculating the electron's wave motion, which determines the shapes, positions, and orientations of the electron's orbital pattern. However, the wave theory equation only applies to a nucleus with one proton and one neutron, i.e., a hydrogen atom. Larger atoms are too diverse to permit this degree of measurement.

Schrödinger equations may be either time-dependent, which considers an electron's unchanging state as a standing wave, or time-independent, which uses wave functions to identify electron densities, including the shapes and sizes of its orbital.

Q 36. Is string theory the same as superstring theory?

The terms are used interchangeably and refer to a theory that quarks and electrons are not the final, indivisible fundamental particles, but all matter is made of almost infinitely small vibrating string-shaped entities. The theory was initiated at CERN, the European agency for nuclear research, by Gabriele Veneziano, and was based on the concept that the gluon, or strong nuclear force, was actually built of the tiny strings that were first thought to be one-dimensional, but later believed to have far more dimensions, ultimately arriving at 10 dimensions. It was then thought that the vibrational patterns of the strings determined what type of particle they formed, with quarks having one type of vibration, and electrons another. Yet another dimension for strings, bringing the total now to 11, was theorized in 1995 by Edward Witten, who further predicted M-theory, that suggests the universe is one of many membrane-shaped forms.

Some theorists propose that all four of the known forces—electromagnetic, strong, weak, and gravity—are made of these vibrating strings, which would help support the long-sought theory that unifies the four forces—the Grand Unified Theory.

Q 37. Why is the Grand Unified Theory called an "interrupted dream"?

Referred to as GUT, the Grand Unified Theory has been called Einstein's interrupted dream, because it is one of his most passionately conceived theories, but has not yet been realized, or proven. Moreover, Einstein himself doubted the uncertainties and inferences of this way of interpreting what he imagined to be an ultimately orderly and logical construction of the components of the universe.

The theory states that at extremely high energy levels, the three gauge forces of the Standard Model— electromagnetism, the strong force, and the weak force—are low-energy manifestations of a single, unified, but much higher energy force. If this were confirmed, it would imply that the three forces had originally been a single force during the early formation of the universe. During the 1960s, experimentation led to formulation of the electroweak force, merging the weak force with electromagnetism. Some physicists are trying to include the fourth force, gravity, into the GUT equation, which would satisfy the concept called "Theory of Everything" (TOE), and unify all four known gauge forces.

A major challenge to proving TOE is the extremely high energy levels required by particle collider experiments to unify the four force particles. This is estimated to be Planck energy, or one billion times 10^{19} electron volts; this is a quadrillion times more energy than the Large Hadron Collider (LHC) can produce.

Q 38. What is supersymmetry and has it been proven?

Supersymmetry is an offshoot or derivative of string theory, and it is why string theory may be referred to as "superstring". Supersymmetry implies that all particles in nature have a larger (or "super") partner, which allows the particle to convert to another form. For example, a quark (which is a fermion, or particle of matter) could convert to a gluon (a boson, the strong nuclear force), without changing structures, thus not violating the laws of conservation of energy and momentum.

If verified, supersymmetry would provide proof of the unification theory that brings gravity into conformity with the other three forces, completing the Standard Model, and confirming Einstein's general theory of relativity, which defines gravity as the bending of space-time by mass (see Q40). Experiments to find supersymmetry particles are continuing at CERN and other high power colliders, but so far, these elusive particles have not yet been discovered.

Q 39. Why is Max Planck called the "father of quantum physics"?

Karl Ernst Ludwig Max Planck was born in Germany in 1858; he later became known as Max as he progressed rapidly through an advanced education. By 1894, he had published new black body radiation theories concerning energy conservation, and began serious focus on how energy travels. Planck determined that light, and all forms of energy, travels not as a continuous flow, as assumed at the time, but in finite bundles, the smallest possible units of energy in nature, which he named "quanta." This was the foundation of quantum physics, and led to his receiving the 1918 Nobel Prize in Physics.

Planck calculated that energy units are always a whole integer, or constant, now called the "Planck constant". He later determined Planck length, Planck energy, Planck time, and Planck temperature; all are the smallest, indivisible units of measurement. Planck was one of the early supporters of Einstein's special theory of relativity in 1905, and helped it gain acceptance in the scientific community. Today, Planck is honored by the existence of 86 Planck scientific institutes worldwide.

Q 40. What is the theory of relativity?

Actually, there are two theories of relativity: We'll define the special theory first, then cover general relativity in a subsequent question.

Albert Einstein published his theory of special relativity in 1905, while he was still working as a patent examiner in Switzerland, conceptualizing theories in his spare time. He was concerned about how light traveled, and the implications of light's fixed velocity, when he discovered a relationship between an object's speed and its mass, especially as the object approaches the speed of light, which is 186,000 miles (300,000 km) per second. Einstein theorized, correctly, that neither matter nor energy can exceed the speed of light, and that the mass of an object becomes infinite as it approaches the speed of light.

Hoping to make his theory understandable, Einstein's famous mind experiment showed that light emitted by a person on a fast moving vehicle or object will always travel at the same speed; the speed of the object and the speed of light are not additive, and the speed of light in a vacuum remains independent of the relative motion of observers.

50 Quantum Physics Questions

Q 41. What exactly does Einstein's $E = mc^2$ mean?

In his follow-up work on special relativity, and the finite nature of the speed of light, Einstein determined that mass and energy are essentially the same, and one can be converted to the other. This eventually led to the development of nuclear energy.

In the formula $E = mc^2$, energy equals mass times c^2, the speed of light squared. Since the speed of light (186,000 miles per second) is a considerably large number, the implication is that a tremendous amount of energy is contained in a small mass. For example, one kilogram (2.2 lbs) of matter contains the energy equivalent of 90-thousand trillion joules, or enough power to light a 100-watt bulb for 28,500 years (if you have the patience and a bulb that would last that long). In a realistic example, less than a kilogram of uranium-235 or plutonium-234 is converted to energy to power a nuclear explosion.

Conversely, a very large amount of energy is needed to create a small amount of mass. Consider how much energy of the early universe was needed to create the massive stars and galaxies we see today, not to mention dark matter and dark energy.

Q 42. How does the general theory of relativity redefine gravity?

Ten years after publishing special relativity and other papers, including one concerning Brownian movement, which confirmed and quantified molecular motion and energy, Einstein published his general theory of relativity. Until this moment, science had accepted Isaac Newton's gravitation theory from the 17th century, which defined gravity as an attractive force between massive objects. But according to general relativity, Einstein introduced time as a fourth dimension, with space and time being interwoven into a single continuum known as space-time—a geometric property of space and time.

In this context, Einstein showed that gravity is actually the bending of space-time by massive objects. Objects on Earth fall, satellites rotate around planets, and planets follow their orbits around stars, according to precise measurements of the bending of space-time. One of the computational findings of general relativity refined measurements of the orbit of Mercury around the Sun, correcting an error in Newtonian physics.

General relativity was finally confirmed—and Einstein's reputation was solidified—on May 29, 1919, during a solar eclipse, when light from a distant star passed close to another star and was "bent" by that star's warping of space-time to the exact degree predicted by general relativity. This effect is put to good use today by astronomers: It's called "gravitational lensing", and it creates both magnification and the ability to determine the mass and density of distant stars and galaxies.

Q 43. Is entanglement really a "spooky action at a distance"?

Entanglement is the effect that causes two particles to influence each other while separatedby a distance. If two subatomic particles have a positive spin of $+\frac{1}{2}$, and you change the spin of one to negative spin $-\frac{1}{2}$, the other particle will instantly convert to $-\frac{1}{2}$. Albert Einstein originated theories of entanglement of physically-separated particles, but questioned the validity of the concept, leading to his "spooky action at a distance" remark. Einstein was not alone in his doubts: Entanglement violates the special relativity law that limits the speed of light, and also violates the principle of uncertainty advanced by Werner Heisenberg and Niels Bohr.

And yet, despite the violation of theoretical principles, the instant interaction of particles does appear to exist. Entanglement is beginning to be realized with tests involving photons, electrons, neutrinos, and atomic nuclei and even larger molecules. Entanglement is being tested for application in communications, quantum radar, as well as for use in quantum computing, as the next question covers.

50 Quantum Physics Questions

Q 44. Is quantum computing real or a hypothetical concept?

Quantum computing is real, but it is in early stages of development. It is considered to have tremendous potential to accelerate computing speed and increase memory. It uses principles of entanglement and superposition, which allows subatomic particles to exist in more than one state. Unlike the computers we use now, which depend on bits of information in the form of a "0" or "1" to indicate no or yes, quantum computers use subatomic particles called "qubits", instead of bits. Qubits, in the form of electrons or photons, can be either 0 or 1 at the same time, thanks to superposition physics, and two qubits can influence each other while separated, thanks to the principle of entanglement. This means changing the state of one qubit, a 0 to a 1, for example, instantly changes its partner particle, and compresses computing time considerably.

Here's how it works: Upon starting to solve a problem, the opening position of the qubits in a quantum computer is a combined state of both 0 and 1. But then, as the computing process reaches a solution, the qubits abandon their superposition duality and assume final sets of strings of 0 and 1 to present a definite result. Quantum computers will not be replacing our laptops and desktops, but applications will be supporting large scale, complex operations on the financial, medical and genetic engineering, and military levels.

50 Quantum Physics Questions

Q 45. Is there more than one universe?

Coming to terms with the established fact that the universe we know (so far) has existed and been expanding for 13.8 billion years, and contains at least 100-billion galaxies, each containing millions and billions of stars, is difficult enough. But what if this universe is not alone, but simply one of many? This is the concept of the multiverse.

One of the most popular concepts is parallel universes, in which there are infinite versions of our universe, each with slight differences. This theory would have an infinite number of planet Earths, and different versions of each of us. A variation of this is called "daughter universes", which exist in infinite versions, but collapse into a single version when measured. Yet another concept has our universe existing as a bubble, in a vast space than contains other bubble universes. Because these universes exist independent of each other, each may have its own unique laws of physics, chemistry, biology, and mathematics.

Another concept, which is less exotic than other theories, is that other universes exist just beyond the reach of our telescopes, and perhaps our universe will expand into neighboring universes.

Q 46. What is the quantum tunneling of an electron?

While we usually imagine an electron to be a particle, it is also a wave function, and as a wave, it has amplitude (height) as well as wavelength. In its orbit around a nucleus, or its travels through a medium such as electricity, the wave function may come up against barriers or obstacles. The amplitude of the wave automatically decreases at an exponential rate, enabling the electron wave function to pass through the obstacle. Assuming the amplitude of the wave function does not shrink to zero, there is a probability that the electron's wave function can be detected on the other side of the barrier. We called this effect "quantum tunneling".

An electron is considered as a wave function. The probability of finding an electron is directly proportional to the square of the amplitude of the wave function. When the wave function of an electron encounters a potential barrier, its amplitude decreases exponentially. For a narrow barrier, the wave amplitude may not become zero after the electron passes through the barrier. Hence, there is a non-zero probability that the electron will be found beyond the barrier. This process is called "quantum tunneling".

Q 47. How does an electron microscope magnify so powerfully?

While a conventional optical microscope magnifies with the photons of visible light, an electron microscope takes advantage of the wave function of electrons. Since the electron wave is 100,000 times shorter than the wavelength of light photons, it gives the electron microscope 100,000 times or more the resolving power. (Louis de Broglie discovered in 1924 that electrons traveling in a vacuum behave like very short wavelength radiation.)

The electron microscope directs a beam of high voltage electrons, up to 100,000 eV, which is generated by a heated metal (tungsten) filament. The beam is accelerated by a positive electrical charge in a vacuum (the positive charge attracts the negatively-charged electron waves). Apertures and magnetic lenses focus the electron wave stream into a narrow, monochromatic, tightly focusedbeam. As the beam is focused onto the specimen or target object, it interacts with its molecules to form an image. Thanks to this technology, previously-invisible particles and viruses can now be seen clearly.

Q 48. How does a PET scan use quantum particles and waves?

Positron emission tomography (PET) scans use short half-life radioactive tracers (typically n-13 ammonia, rubidium-82, technetium-99, and fluorine-18) to create medical images of metabolic processes within tissues and organs, and measure conditions and changes, by applying quantum mechanics technology. The scans are most often used to measure the metabolic rate of cancer cells to reveal decreased blood flow to the heart, and to detect rates of glucose utilization in various parts of the brain. It is also used in evaluating patients with Alzheimer's disease, depression, epilepsy, and Parkinson's disease. PET scans evaluate conditions at the cellular level; MRIs and CT scans cannot.

The nuclei of the radioactive tracer emit positrons, which interact with electrons in the protein atoms of the body part being studied. The negative electron and positively-charged positron annihilate each other, producing gamma rays, which travel in 180 degree opposing directions. The detector ring of the scanner detects and records the location and energy of the gamma rays, leading to the construction of detailed 3D images of the subject organ, tumor, or tissue. For example, a PET scan can identify where the myocardium (heart muscle) may be damaged due to reduced blood flow.

Q 49. What is Pauli's exclusion principle and why is it important in chemistry?

The exclusion principle was proposed by Austrian physicist Wolfgang Pauli in 1925 to describe the behavior of electrons. Exclusion states that no two electrons can occupy the same quantum position in an atom or molecule. Electron positions and states are defined by four quantum numbers—n, l, m_l, and m_s—the last, m_s, symbolizing the electron's half integer spin: $+\frac{1}{2}$ or $-\frac{1}{2}$. Here is an example: The two electrons in the first orbital shell of a helium atom have the same first three numbers ($n=1$, $l=0$, and $m_l=0$), but m_s cannot be the same for both, so one electron's spin is $+\frac{1}{2}$, and the other's spin is $-\frac{1}{2}$.

In 1940, Pauli extended the exclusion principle to include all fermions, which have $\frac{1}{2}$ integer spin. Since bosons (particles that carry force) have full integer spins ($+1$, -1), exclusion does not apply; this permits photons of light to occupy the same quantum position in a laser beam.

The Pauli exclusion principle is applied in chemistry to identify the electron shell structure of atoms, by predicting the atoms that can share electrons in chemical bonds.

50 Quantum Physics Questions

Q 50. What are Quantum Electrodynamics and Quantum Chromodynamics?

This final question returns to concepts we've covered, to provide definitive clarification of what goes on at the quantum level. QED and QCD both describe interactions, but of very different levels of force, mediated by bosons, which are massless particles.

Quantum Electrodynamics (QED) is the quantum field theory that defines the electromagnetic force and the emission and absorption of photons—the particle that transmits or mediates the electromagnetic force.

Quantum Chromodynamics (QCD) was developed as an analogy to QED, and describes the strong interaction between quarks and gluons; QCD defines the nature and characteristics of quarks and gluons within atoms, by acting both on the fundamental quarks, and composite particles: Protons, neutrons, and unstable mesons. The gluon carries the strong force that holds quarks, protons, neutrons, and atomic nuclei together.

QED has one type of charge, either positive-electric or negative-electric. In contrast, the word "chromodynamics" in QCD refers to colors, a quantum number assigned to quarks and gluons, symbolically referred to as red, green, and blue (these are symbolic—there are no actual colors involved). Similarly, gluon particles also carry symbolic colors; there are a total of eight colors that enable the gluons to interact with the quarks.

Conclusion

I hope that your questions about the mysteries and scientific advances in quantum physics have been well-answered, and that your curiosity has been stimulated to look further to explore the issues within quantum physics that interest you most. In the following reference section, you will find links to the many publications that were sourced to provide accurate and up-to-date information on the historic and contemporary theories, established scientific laws, and new findings in quantum physics.

You may have been inspired, as I have been for some time, by the courageous, insightful pioneering scientists who paved the way for our understanding of the universe, especially at the invisible, subatomic level, with theories and experiments. We owe a great debt of gratitude to Max Planck, Albert Einstein, Marie and Pierre Curie, Louis de Broglie, Edwin Schrödinger, Niels Bohr, Werner Heisenberg, Wolfgang Pauli, and many others—I encourage you to learn more about them.

If you enjoyed this book as much as I think you will, please take a moment to give the book a good rating, to encourage others to share in the quantum physics insights the book provides.

All the best in your continuing quest for knowledge,

Donald B. Grey

Bluesource And Friends

This book is brought to you by Bluesource And Friends, a happy book publishing company.

Our motto is **"Happiness Within Pages"**

We promise to deliver amazing value to readers with our books.

We also appreciate honest book reviews from our readers.

Connect with us on our Facebook page www.facebook.com/bluesourceandfriends and stay tuned to our latest book promotions and free giveaways.

References

Atteberry, J. (2020). What exactly is the Higgs boson? *Science - How Stuff Works.*https://science.howstuffworks.com/higgs-boson.htm

Allpress, K. (2018, February 7). Why do electrons in an atom keep a distance from the protons if opposite charges attract? *Quora.* https://www.quora.com/Why-do-electrons-in-an-atom-keep-a-distance-from-the-protons-if-opposite-charges-attract-Why-dont-electrons-crash-into-the-nucleus

Chemcool Dictionary. (2017). Definition of the Schrödinger equation. https://www.chemicool.com/definition/schrodinger_equation.html

Chemistry Libre Texts. (2019, July 1). Radioactivity: alpha, beta, and gamma decay. https://chem.libretexts.org/Courses/can/intro/17%3A_Radioactivity_and_Nuclear_Chemistry/17.03%3A_Types_of_Radioactivity%3A_Alpha%2C_Beta%2C_and_Gamma_Decay

Conn, R. (2019, July 5). Nuclear fusion. *Britannica.* https://www.britannica.com/science/nuclear-fusion

Conner, N. (2019, December 14). What is radioactive half-life - physical half-life - definition. *Radiation*

50 Quantum Physics Questions

*Dosimetry.*https://www.radiation-dosimetry.org/what-is-radioactive-half-life-physical-half-life-definition/

DOE Science News Source. (2020, August 13). UChicago scientists discover way to make quantum states last 10,000 times longer. https://www.newswise.com/doescience/?article_id=736376&returnurl=aHR0cHM6Ly93d3cubmV3c3dpc2UuY29tL2FydGljbGVzL2xpc3Q=&sc=sphn

E=mc^2. (2020) E=mc^{2}Solving the equation. https://www.emc2-explained.info/Emc2/Equation.htm

Editors of Encyclopedia Britannica. (1998, December 26). De Broglie wave. https://www.britannica.com/science/complementarity-principle

Editors of Encyclopedia Britannica. (2018, January 9). Spin, atomic physics. https://www.britannica.com/science/spin-atomic-physics

Editors of Encyclopedia Britannica. (2020, May 27).Uncertainty principle. https://www.britannica.com/science/uncertainty-principle

Fermilab Visual Media Services. (2020). Grand Unified Theory: GUTs and TOEs. https://duckduckgo.com/?t=ffab&q=grand+unified+theory&atb=v215-1&ia=videos&iax=videos&iai=https%3A%2F%2Fwww.youtube.com%2Fwatch%3Fv%3D9LGBo7dLgYk

Fisher, L. (2020). What's the distance from the nucleus to an electron? *Science Focus*.https://www.sciencefocus.com/science/whats-the-distance-from-a-nucleus-to-an-electron/

Helmenstine, A-M. (2019, December 8). Nuclear fission definition and examples. *Thought.co*. https://www.thoughtco.com/nuclear-fission-definition-and-examples-4065372

Helmenstine, A-M. (2019, February 02). Pauli exclusion principle definition. *Thought.co*.https://www.thoughtco.com/definition-of-pauli-exclusion-principle-605486

Krans, B., Adcox, M. (2018, September 17). What is a PET scan? *Healthline*.https://www.healthline.com/health/pet-scan

Lim, A. (2019, August 9). Henri Becquerel and the serendipitous discovery of radioactivity. *Thought.co*. https://www.thoughtco.com/henri-becquerel-radioactivity-4570960

Live Science Staff. (2014, June 20). What is antimatter? https://www.livescience.com/32387-what-is-antimatter.html

Milunski, M. (2013, November 22). What isotope is used in the injection for a PET scan? *Healthtap*.https://www.healthtap.com/questions/22908-what-isotope-is-used-in-the-injection-for-a-pet-scan/

NASA Science. (2020). Dark energy, dark matter. https://science.nasa.gov/astrophysics/focus-areas/what-is-dark-energy

Obodovskiy, I. (2019). Gamma radiation: nuclei and nuclear radiation. *Science Direct.*https://www.sciencedirect.com/topics/physics-and-astronomy/gamma-radiation

Office of Nuclear Energy. (2020). Nuclear fuel facts: uranium. https://www.energy.gov/ne/fuel-cycle-technologies/uranium-management-and-policy/nuclear-fuel-facts-uranium

O'Neill, I. (2017, May 29). How a total solar eclipse proved Einstein right about relativity. *Space.com.* https://www.space.com/37018-solar-eclipse-proved-einstein-relativity-right.html

PhysicsNet. (2020). Particles, antiparticles, and photons. http://physicsnet.co.uk/a-level-physics-as-a2/particles-radiation/particles-antiparticles-photons/

Ptable. (2020). Dynamic periodic table of the elements. https://ptable.com/

Redd, N.T. (2017, November 7). Einstein's theory of general relativity. *Space.com.*https://www.space.com/17661-theory-general-relativity.html

Scientific American. (1999, October 21). What exactly is the 'spin' of subatomic particles? https://www.scientificamerican.com/article/what-exactly-is-the-spin/

Sutton, C. (2017, April 17). Standard Model. *Britannica,*https://www.britannica.com/science/Standard-Model

50 Quantum Physics Questions

VA.gov. (2020). What is an electron microscope and how does it work?
https://www.va.gov/DIAGNOSTICEM/What_Is_El
ectron_Microscopy_and_How_Does_It_Work.asp

Wikipedia. (2020). Cosmic microwave background.
https://en.wikipedia.org/wiki/Cosmic_microwave_bac
kground

Wikipedia. (2020). Gamma rays.
https://en.wikipedia.org/wiki/Gamma_ray

Wikipedia. (2020) Pion. https://en.wikipedia.org/wiki/Pion

Wikipedia. (2020). W and Z bosons.
https://en.wikipedia.org/wiki/W_and_Z_bosons

Zimmerman-Jones, A. (2019). Laws of thermodynamics.
*Thought.Co.*https://www.thoughtco.com/laws-of-
thermodynamics-p3-2699420

Zimmerman-Jones, A. (2019, July 3). Wave particle duality and
how it works.
*Thought.Co.*https://www.thoughtco.com/wave-particle-
duality-2699037

Zimmerman-Jones, A. (2019, October 15). What you need to
know about the weak force.
*ThoughtCo.*https://www.thoughtco.com/weak-force-
2699335

Zimmerman-Jones, A. (2019, January 19). The 4 fundamental
forces of physics.
*ThoughtCo.*https://www.thoughtco.com/what-are-
fundamental-forces-of-physics-2699070